Beware! These Animals Are Poison

BARBARA BRENNER

Beware! These

Animals Are Poison

Illustrated by Jim Spanfeller

Coward, McCann & Geoghegan, Inc. • *New York*

Text copyright © 1979 by Barbara Brenner

Illustrations copyright © 1979 by Jim Spanfeller

All rights reserved. This book, or parts thereof, may not be reproduced in any form without permission in writing from the publishers. Published simultaneously in Canada by Longman Canada Limited, Toronto.

Library of Congress Cataloging in Publication Data

Brenner, Barbara.   Beware! These animals are poison.

   SUMMARY: *Discusses various facts about poisonous animals with emphasis on how they deliver their venom to other animals as a means of survival.*

[1. Poisonous animals]   I.   Spanfeller, James J., date. II. Title.
QL100.B73   591.6′9   77-27591
ISBN 0-698-20438-7

Designed by Lynn Braswell
Printed in the United States of America

Acknowledgments

One of the fringe benefits of writing is that it's always a learning experience for author as well as reader. I want to thank George Foley, Department of Herpetology, American Museum of Natural History, New York City; Alice Gray, Department of Entomology, American Museum of Natural History, New York City; and George Ruggieri, Director, New York Aquarium, Coney Island, New York. They made interesting and valuable corrections and additions and contributed additional perspective on the subject of poisons and poisonous animals.

Contents

1 · A Matter of Life and Death 11
  WHY DO ANIMALS CARRY POISON? · WHAT DOES IT DO? · WHAT IS THE POISON MADE OF?

2 · Tentacles 15
  PORTUGUESE MAN-OF-WAR · SEA WASPS · SEA ANEMONE

3 · The Sting 21
  BEES · WASPS · SCORPIONS

4 · Fangs 27
  SNAKES · SPIDERS

5 · **Spines, Daggers and Other Weapons**   *37*
     SCORPION FISH · STINGRAY · GIANT
     WATER BUG

6 · **Body Poison**   *41*
     PUFFER FISH · SEA URCHINS · TOADS ·
     FROGS · NEWTS · MILLIPEDES

7 · **Tooth and Claw**   *51*
     CONE SHELL · SHORT-TAILED SHREW ·
     DUCK-BILLED PLATYPUS

8 · **The Smallest and the Deadliest**   *55*
     ZOANTHIDS

9 · **A Final Word**   *57*
     RESPECT FOR POISON CREATURES · HOW NOT TO
     BE AN ALIEN PRESENCE

poison (poi′zən) 1. Any substance which causes injury or death. 2. A chemical which can result in pain, swelling, dizziness, nausea, paralysis, shortness of breath, heart failure, death.

venom (ven′əm) The poison of certain animals, such as some snakes, fish, spiders, insects. Usually associated with those poisonous animals which can deliver poison by a sting, bite, or other means.

1 · A Matter of Life and Death

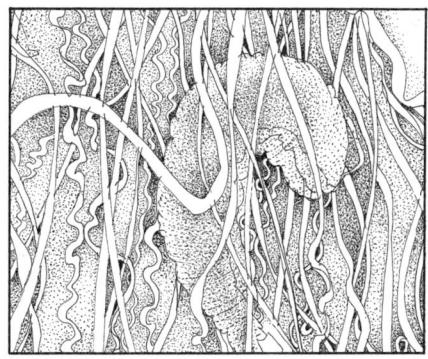

Poison. Venom. They're powerful words. The language of violence. When someone says *venom*, we think *hate*. When we hear *poison*, we think *death*.

But in nature these words have a different meaning. For certain animals, they can mean *life*. Armed with their poison weapons, venomous creatures can catch and hold their prey. Or they can drive away enemies. Poison enables them to stay alive—to have young and in that way make sure that their species survive.

Isn't some animal poison deadly to people, too? Yes. But its main purpose is not to "get" us humans. Poisonous animals aren't "bad." Or mean. Or cruel. We have to think of their poison in another way. Poisonous animals are simply creatures that have developed a special kind of survival kit.

There are hundreds of poisonous animals. A few of them are giants, like one species of stingray which weighs 700 pounds (315 kilos). But most poison creatures are small, even tiny. Like the zoanthids on page 55 in this book.

All poisonous animals, large or small, have one thing in common. *Their bodies make poison.*

The poison is made up of chemicals. Some poisons are simple—one or two chemicals. Others contain a mixture of many chemical substances.

☠ Some of the simplest animals make the most complicated poisons.

What, exactly, can these chemicals do? They can destroy blood cells, dissolve tissue, interfere with breathing, or make a heart stop beating. They can paralyze or kill in an instant. Each animal's poison is its own. But some animals which are not at all alike have similar poisons.

☠ The puffer fish and the California newt, for instance, have the same chemical in their poisons.

Animals deliver venom in a variety of ways. Tentacles, stings, fangs, spines, teeth, beaks, claws—an endless number of weapons do the job. And some animals poison only if they are eaten.

Let's take a look at some of the best known—and some of the least known—poisoners.

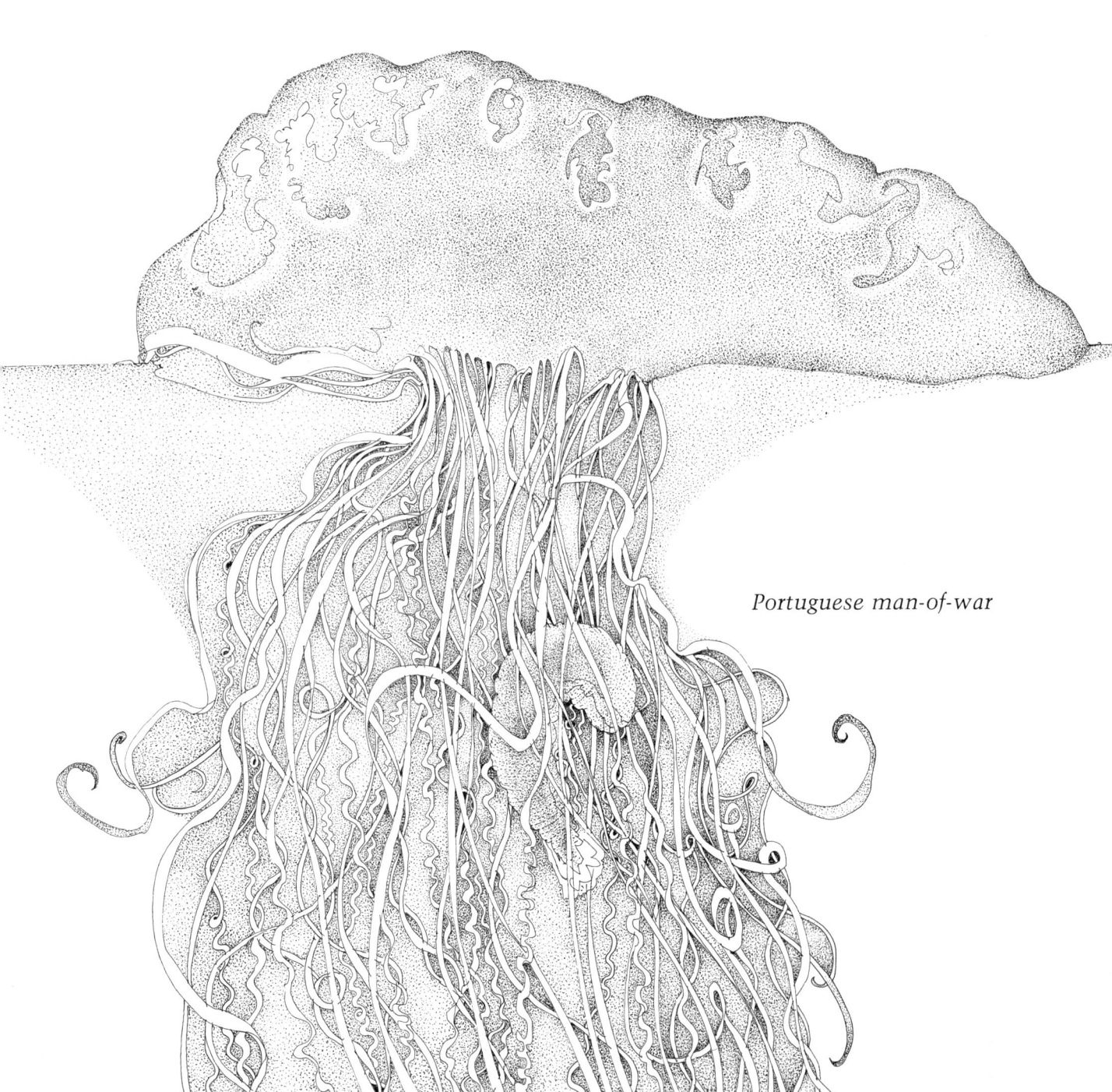

Portuguese man-of-war

2 · Tentacles

PORTUGUESE MAN-OF-WAR

The Portuguese man-of-war looks like a strangely shaped balloon with strings. But it's really a colony of small sea animals. The "balloon" part keeps it afloat. The "strings" that trail under water are tentacles. Sometimes they're as much as 50 feet (14.5 meters) long. And they're covered with bead-like stinging cells. When the tentacles touch possible food, each "bead" puts out a thread with a poison barb on it. The barb sinks into the shrimp or other small sea creature. It's quick death for the prey and a meal for the man-of-war.

Although the poison tentacles are used to get food, they will sting anything they touch, which is why swimmers and scuba divers are urged to stay clear of these floating menaces.

However, there are a few creatures which don't seem to feel the painful venom of the man-of-war. One is the loggerhead turtle. These turtles have been seen feeding on a group of Portuguese men-of-war with their eyes swollen shut from the stings.

SEA WASPS

Sea wasps are transparent. And some of them are no bigger than a grape. The bodies of these jellyfish are mostly water. Leave one in the sun for a while and it will become a blob of liquid. They're almost not there at all. But get in the way of their stinging tentacles and you could be dead in minutes!

Sea wasp is only one of their names. Another is box jelly. They're also called fire medusae. Whatever you call

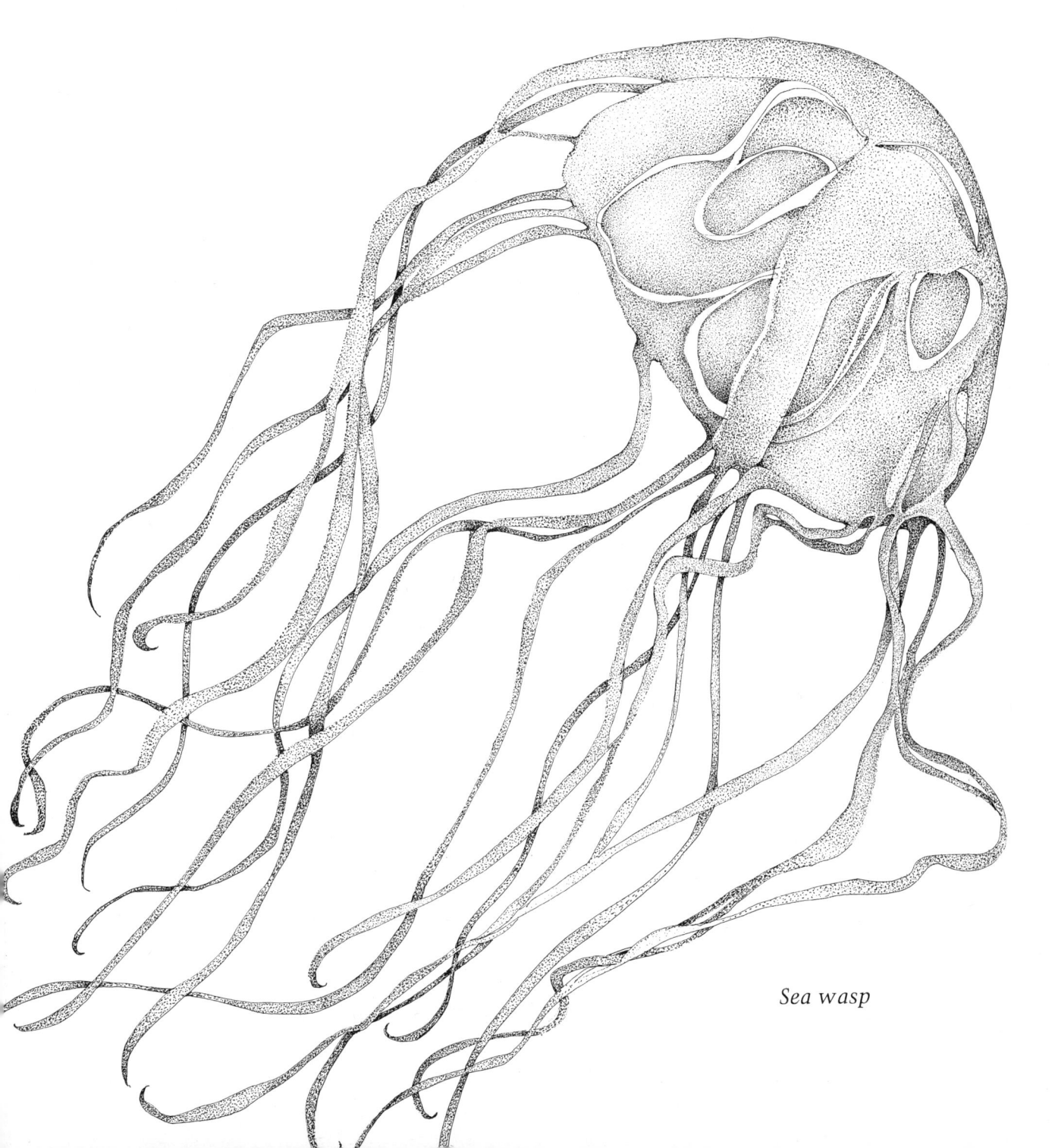

Sea wasp

them, these little creatures are the deadliest jellyfish in the world.

SEA ANEMONE

It's called *flower* of the sea, but it's really an animal. A very simple animal. Just a bag, a "foot," and a ring of tentacles. Like its relatives, the jellyfish, the sea anemone gets its food by "feeling" the water with its tentacles. When they touch something, the tentacles go into action. Like the jellyfish, they also have stinging cells. Some of them shoot poison into the victim. Other tentacles wrap around and hold it. Still other tentacles stick to the prey and pull it toward the mouth of the anemone.

Is any small sea creature safe from this venomous "flower"? The clownfish is. But *only* if it is the clownfish which lives in that particular anemone. One or more clownfish may live in an anemone unharmed. But if a strange clownfish comes near, it is immediately poisoned and eaten.

☠ Scientists still haven't learned exactly how the anemone "knows" its own clownfish, or how these friendships start.

Sea anemone with clownfish

3 · The Sting

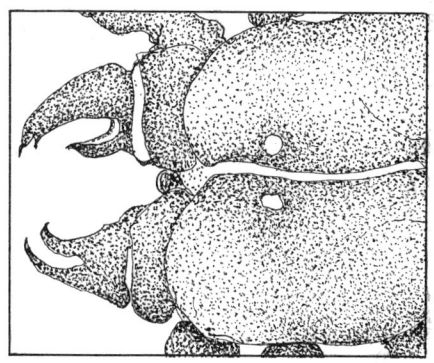

BEES

Bees are poisonous insects. The chemical in a bee's sting (which most people call a stinger) deserves that name as much as any other animal venom.

Only the female bee secretes poison. Her sting is a sharp, hollow needle which can fill with venom. She uses it to protect the hive and the queen bee. The bumblebee, for instance, will guard the hive against anything from a robber

Bumblebee

Honeybee

fly to a skunk. In action, she'll bite as well as sting. Sometimes she rolls onto her back, her jaws open and her sting sticking up. In this position, she may even squirt venom into the air!

An angry bumblebee can sting many times. Not so the honeybee. She has a small barb at the end of her sting. It can't be pulled out of a victim. And the honeybee can't live without it. So when this bee uses her poison, she loses her life.

☠ Then why does the honeybee do it? Because her poison works for the survival of the group—the queen and the young bees.

WASPS

Like bees, wasps carry venom in their stings. Many species of wasps live in groups, in some form of nest or "house." In these colonies, the females use their venom to defend the home. They may even "gang up" on an intruder and sting in a group.

Other wasps poison for another purpose. One is a wasp known as the tarantula hawk. The female's sting is used mainly against tarantulas, which are large, venomous spiders that are fierce fighters. The tarantula hawk almost always wins out over the tarantula. After she stings and paralyzes the spider, she buries it alive in an underground nest. She lays a single egg in the nest. When the egg hatches, the young wasp feeds and grows on the tarantula, which is slowly eaten alive.

☠ The wasp never paralyzes more prey than the young one can eat. Or fewer. Always the right amount.

SCORPIONS

The scorpion is related to the spider. But it looks more like its distant relative, the lobster. There are 650 species of scorpions. All of them have a sting and venom glands in their tails. But a scorpion will use its poison to get food only if its powerful claws fail to do the job of catching and holding prey. When that happens, the scorpion will bring its tail up over its back and *zap* the victim with a dose of venom. Species with small claws sting more frequently; they need their poison to subdue prey.

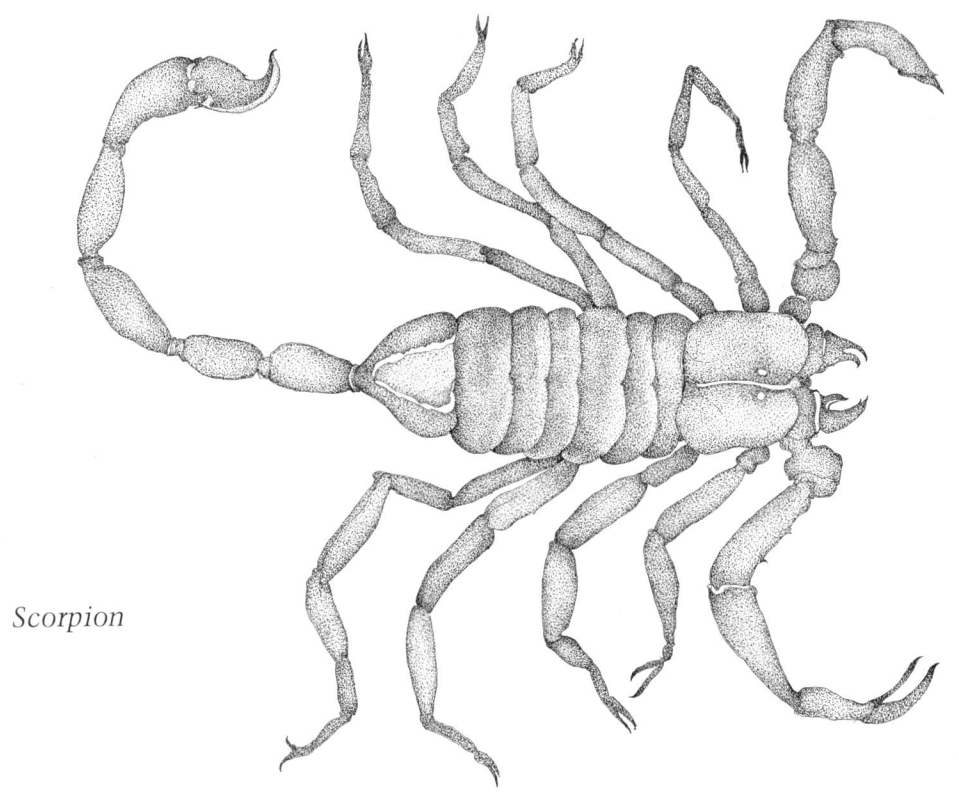

Scorpion

    Scorpions will also use their venom for defense, even against humans. In desert places, a scorpion will sometimes crawl into a camper's pack. The sting of two out of the twelve species in the United States is dangerous.

☠ Good advice for campers, especially in the desert: Shake clothes out before you wear them.

4 · Fangs

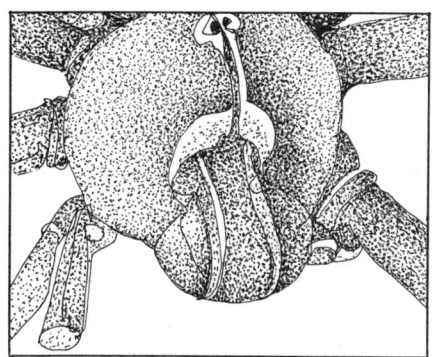

SNAKES

The poison power of certain snakes has been known for a long time. In ancient Egypt death by snakebite was a punishment for murder. And Cleopatra is supposed to have killed herself by allowing a cobra to bite her. During the war between Rome and Carthage, the Carthaginians threw pots of live snakes onto the decks of the Roman ships. Venomous snakes have been used to protect holy places, to bring good luck, even to make rain.

☠   The Hopi Indians still use live rattlesnakes in their rain ceremony.

Not all snakes are venomous. The ones that are have special glands where the poison is produced. It is delivered through the snake's fangs (hollow teeth) or by means of enlarged or grooved teeth. A snake with fangs injects venom into its victim when it strikes. A snake with enlarged or grooved teeth bites and usually chews on its prey while venom runs into the entry wound. There are also several species of snakes which spit or spray venom. The "spitting" cobras of Africa defend themselves by spraying venom at the eyes of an enemy. They can spit for distances of up to 9 feet (2.75 meters) and have been known to blind a victim.

☠   Speaking of spit, the closest thing to venom in human beings is saliva—but it's not poisonous.

Many snakes have a bite that is deadly to humans. But considering the number of poisonous snakes in the world, relatively few people die from snakebite (about a dozen a year in the United States). Most snakes will retreat or hide rather than bite a person. They use their poison mainly to get food. Venom makes it possible for a snake to catch its meal quickly, without wasting energy, and without having

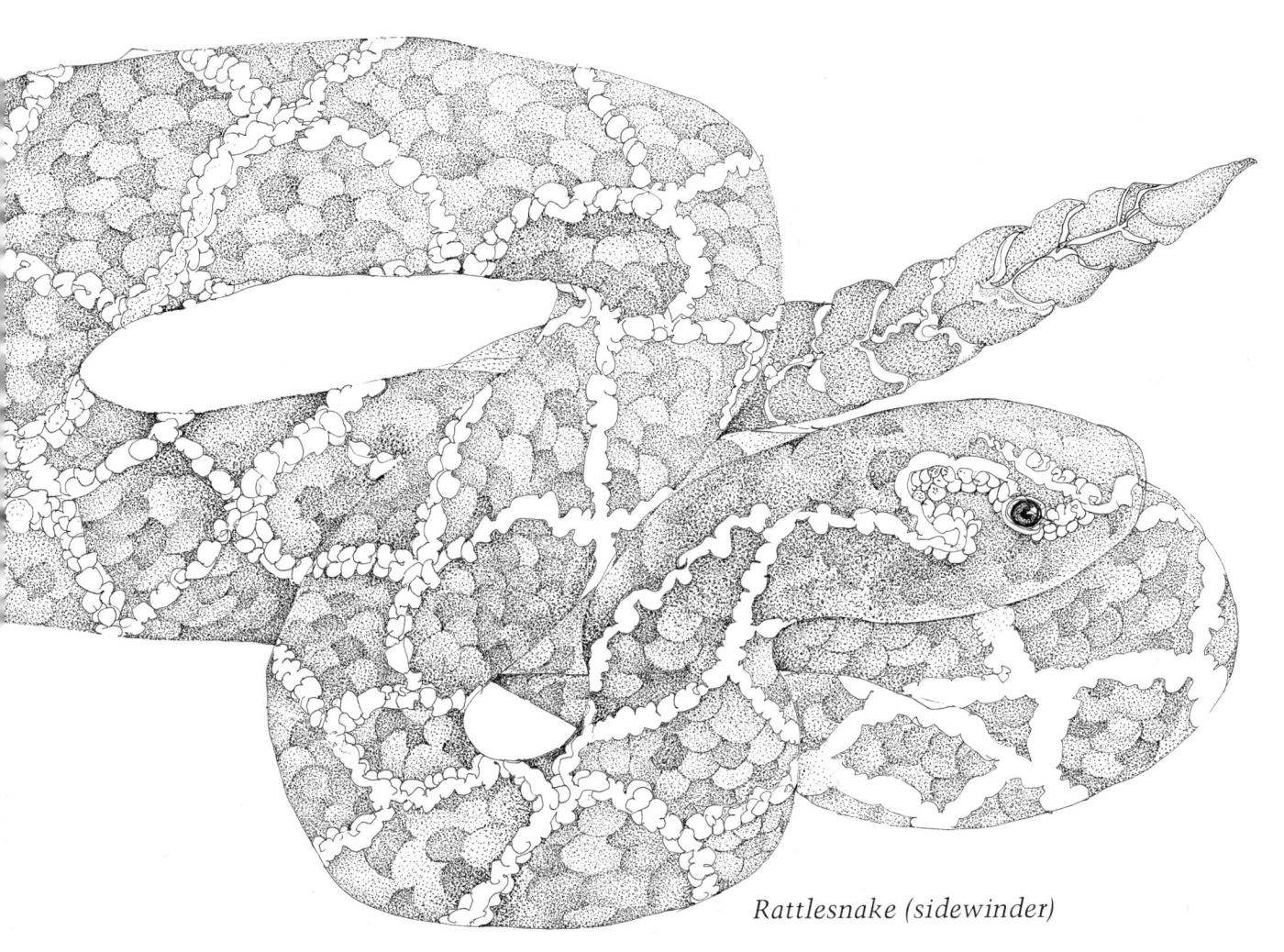
Rattlesnake (sidewinder)

to stay close to its prey, which could be dangerous.

The strength of a snake's poison may have little to do with its size. Baby snakes can be just as venomous as their parents. They need to be. As soon as they're born they are on their own and have to get food for themselves.

The sidewinder is one of fifteen species of rattlesnakes found in the United States and Canada. All rattlesnakes have fangs in the front of the upper jaw. If a rattler loses a fang, it is replaced almost immediately by another one.

Boomslang means *tree snake* in Afrikaans. This snake's fangs are in the rear of the upper jaw. Snakes of this type are called rear-fanged snakes.

☠ It takes only .0002 milligrams (which is less than a millionth of an ounce) of boomslang venom to kill a pigeon.

The amount of venom that a snake injects when it strikes varies greatly. An angry snake may deliver more venom than one that's giving a warning bite. And biting uses up venom. The bushmaster that has just killed a rat will have less venom with which to strike a new target.

When a cottonmouth water moccasin strikes a mouse, here's what usually happens: The mouse may run a short distance, then stop. It rubs its body where the bite is. Then,

Boomslang

Bushmaster

for a few minutes, it remains quiet. It is in shock. Now the wound may begin to ooze. At this point the mouse acts as if nothing had happened. Suddenly it grows very still. Then it falls over, dead. The snake will now move cautiously to the dead mouse and begin to eat it.

Scientists "milk" poisonous snakes to study their venom. One thing they've discovered is that no two venoms are exactly alike. Cobra venom, for instance, acts mainly on the nervous system. Rattlesnake venom attacks blood and tissue.

☠ It takes ten healthy timber rattlesnakes to produce 1 teaspoon (5 milliliters) of venom.

What does venom look like? Most venoms are pale yellow liquids. If you were to drink a small amount by mistake, it probably wouldn't hurt you. But if the same amount were injected into your bloodstream, it could be fatal.

In some parts of the world snake venom is being used to make medicine for treating human diseases. It is also the main ingredient in the serum for snakebite.

☠ Snake venom, dried to a powder, will keep its strength for as long as fifty years.

SPIDERS

Most spiders are mildly venomous. They carry just enough poison to "knock out" their prey, but not enough to hurt a person. The knockout drops are delivered by the spider's fangs, which are hollow tubes in the front part of its body. Because the venom paralyzes, a spider can eat without fear that its meal will crawl or fly away. And spiders have a strong drive to eat. So any creature caught in a spider's web can expect the worst. First it will be paralyzed by a bite from the spider's fangs. Then it may be eaten right away. Or it may be wrapped in webbing like a mummy and stored alive for future use. At mealtime its body will be sucked dry. And finally, when it is a dry shell, it will be tossed out of the web when the spider cleans house.

☠ A spider will even eat its own leg if the leg should come off in an accident.

The female black widow is famous for being one of the few spiders harmful to people. Not so famous, but just as dangerous, is the brown recluse. A bite from this spider's fangs leaves a nasty, rotting sore which may take a long time to heal.

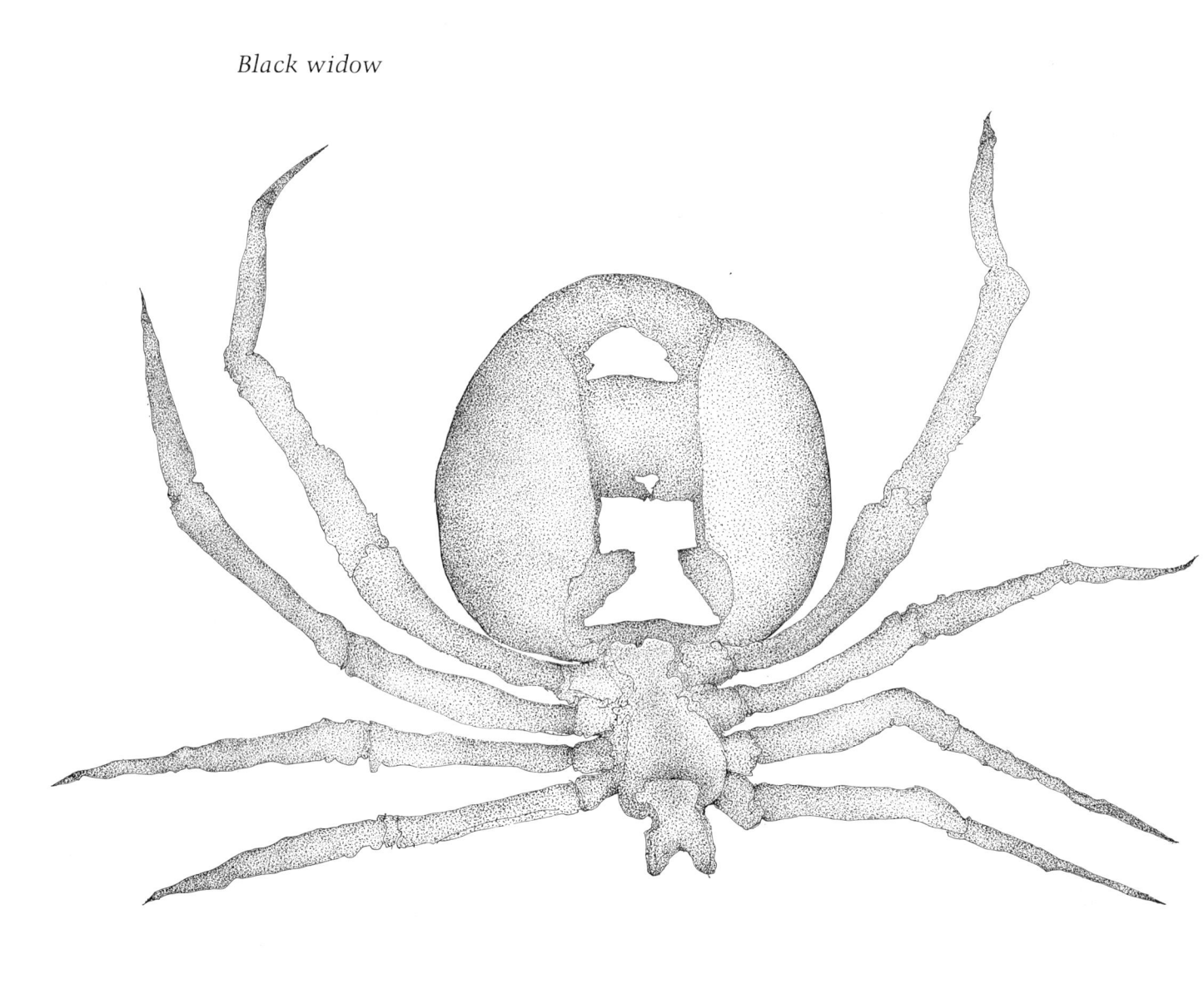

Black widow

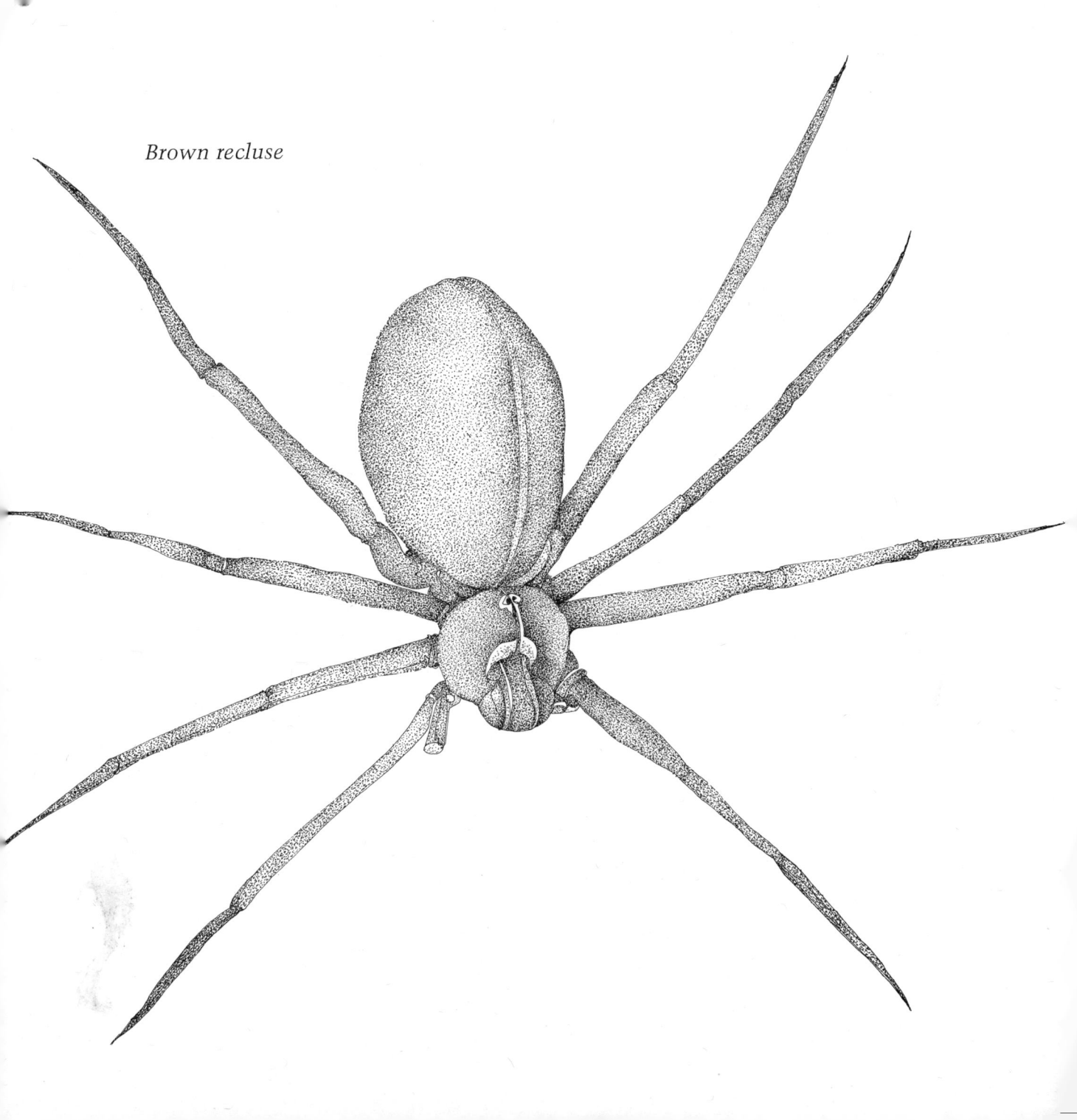

Brown recluse

Why would one of these spiders bite a person? Does a finger or leg look like some giant piece of food? Not really. But spiders hide in dark places—like old barns, outhouses, and abandoned woodpiles. A human invader who disturbs the spider in its hideout could provoke the stabbing thrust which is known as a spider "bite."

☠ Fatal spider bites are rare. There have been less than one hundred known deaths from black widow bites in the last two hundred years.

5 · Spines, Daggers and Other Weapons

SCORPION FISH

Scorpion fish is the name for a group of venomous fishes. One species is called a zebra fish because of its stripes. It is very beautiful. But *beautiful* and *ugly* are words that have different meanings in the world of poison. In fact, beautiful colors may be a warning to other creatures to stay away. The zebra fish carries deadly venom in spines on its back and fins. They are used for defense. But this fish will not

Stonefish

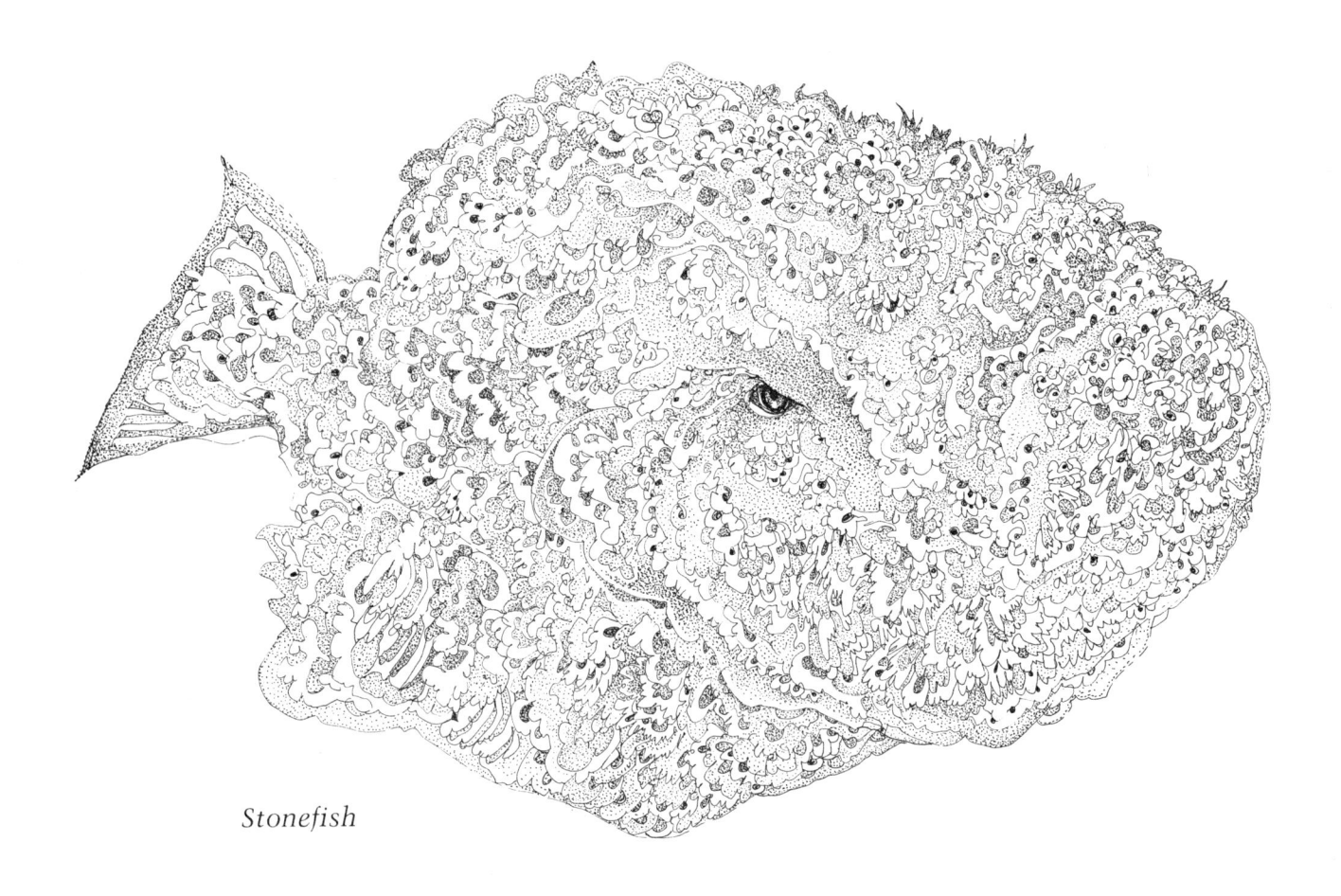

only defend its territory, it will attack to protect itself from what it senses as danger.

What is that lying on the bottom of the tide pool? It

looks like a rock covered with slime. It looks like there's seaweed growing on it. But it's a living thing—a stonefish—lying in wait for a small fish to swim by. That's when the stonefish will make its move. It happens so fast your eye may miss it. But the small fish disappears.

The stonefish is the most venomous of all the scorpion fish. It has 13 spines, an unlucky number for anything that swims too close, or brushes against this deadly "rock."

STINGRAY

It's called a stingray. But it doesn't sting. It *stabs*, with a saw-toothed poison dagger located in its tail. And this weapon knows no difference between a fish and a human foot. A stingray will lie at the bottom of a shallow tide pool or bay, half buried in sand. If it is disturbed it will burst out of its hiding place. The tail will lash in all directions like a whip. Anything in the way will be cut. The dagger part may even stay stuck in the victim. If this happens, the stingray will grow a new one.

There are about one hundred different kinds of venomous stingrays. But many rays are not venomous and use their tails strictly for swimming.

Giant water bug

GIANT WATER BUG

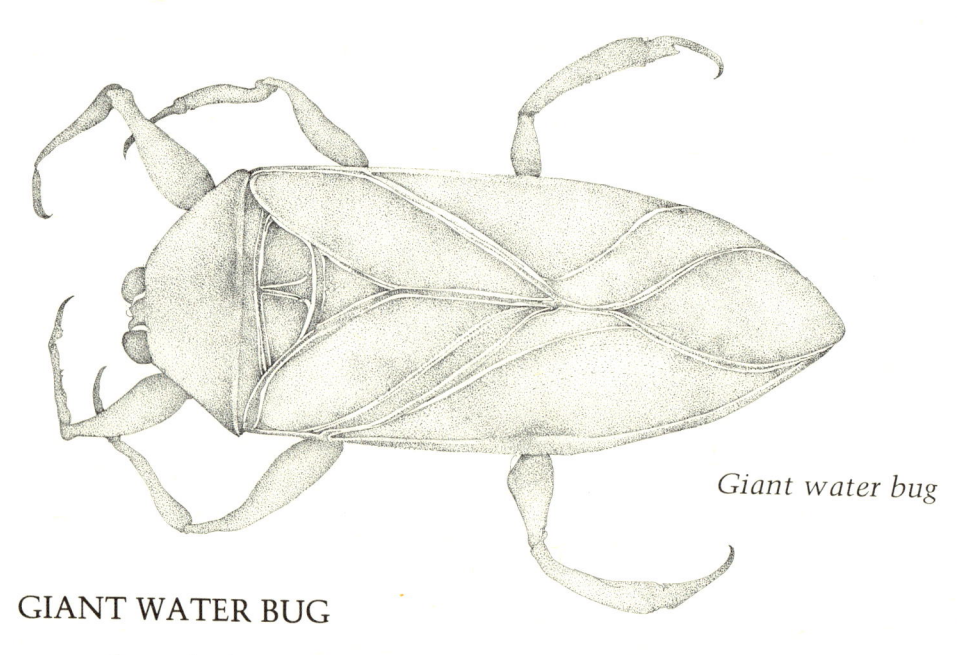

Water snakes, fish, frogs, *beware!* Here comes the giant water bug—2 inches long (5 centimeters). Powerful back legs for swimming. Front legs for grabbing and holding. And a beak loaded with poison. The giant water bug attacks with its beak first, stunning its prey with venom. Then it uses the beak like a straw. The venom turns even a big bullfrog into liquid "soup" for the bug to drink.

☠ The telltale mark of a giant water bug's presence is the empty skin of a frog or fish floating on a pond.

6 · Body Poison

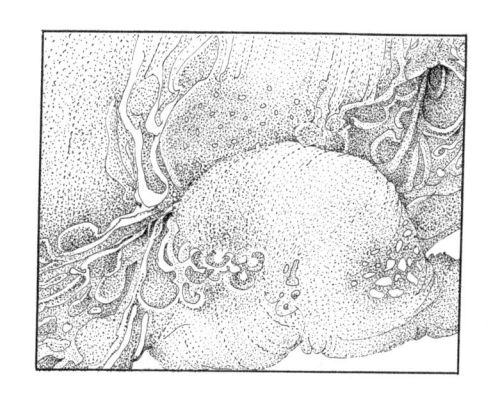

PUFFER FISH

Uk! Uk! Uk! That gurgling sound is a puffer fish blowing itself up with water. That's to make it look fierce and too big for another fish to swallow. It's the puffer's defense against being eaten. But if that doesn't work, some puffer fish have another weapon. They have poison in their skin and other organs. Not much is known about what this poison does to other fish. But we do know that people have died from eating puffers.

☠ The Polynesians call this fish *maki-maki*. Deadly death.

What happens when two poisonous sea creatures meet? Do they poison each other? Is it a standoff? Let's see.

A puffer fish heads for a sea urchin, a small relative of the corals and sea anemones. Not all sea urchins are venomous. But this one is. Each one of its porcupine-like quills has a poison tip. This sea urchin lives on a coral reef. Right now it's sitting on the bottom of a shallow tide pool, chewing seaweed. Its tube feet are planted firmly in the sand. Its poison spines are ready for action.

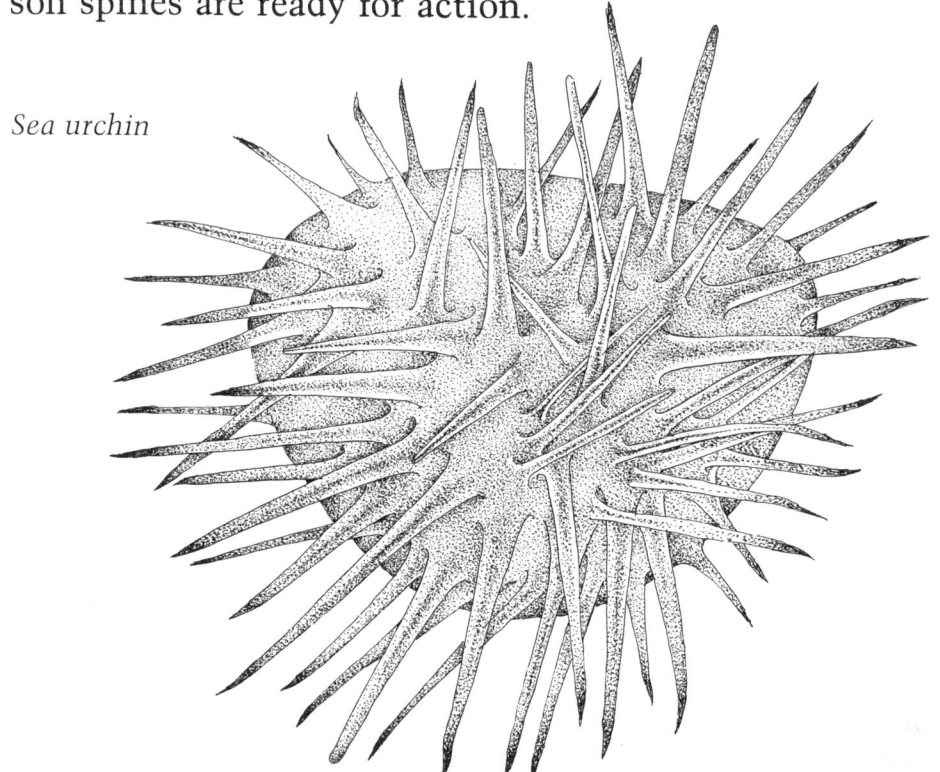

Sea urchin

Puffer fish

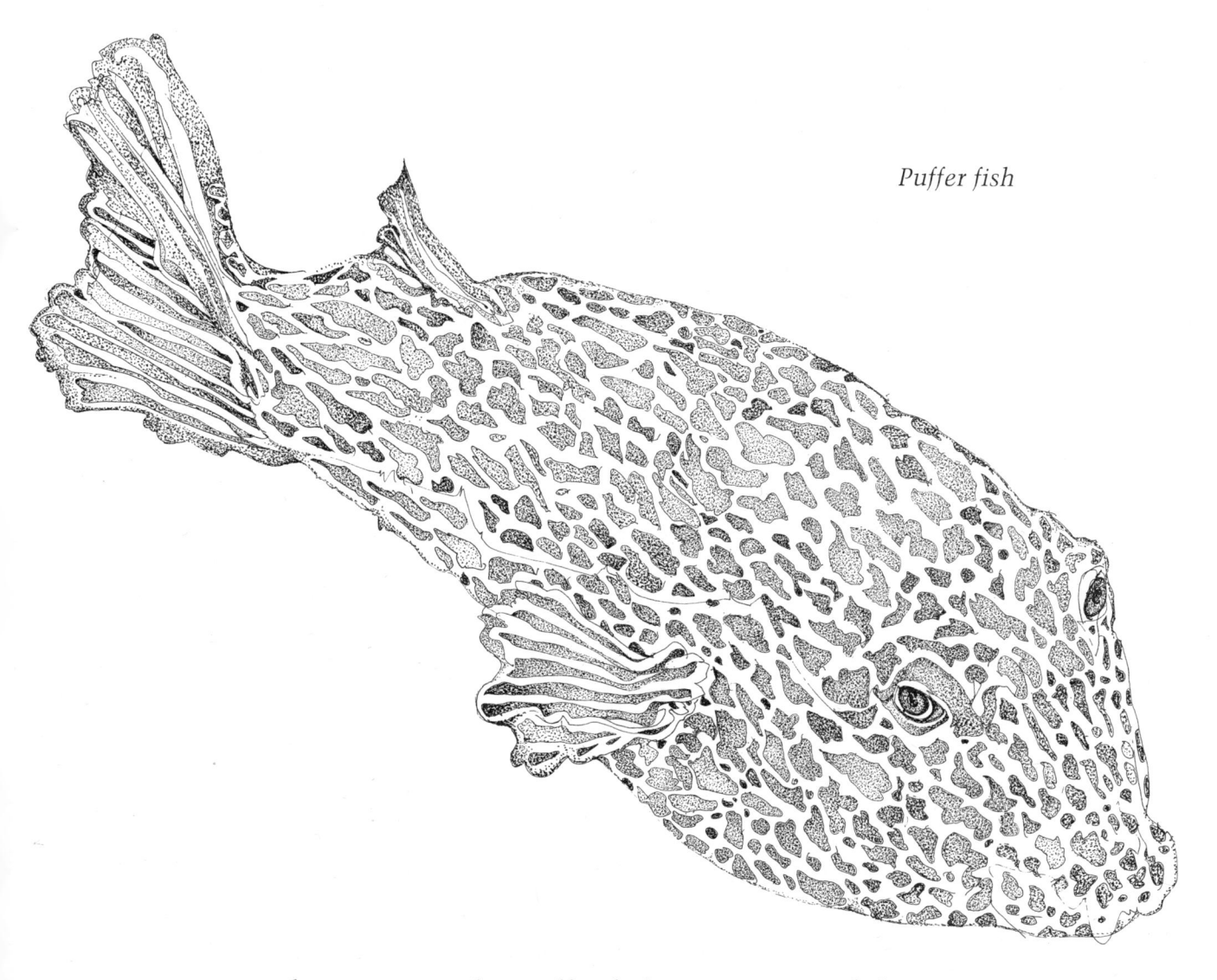

Now here comes the puffer fish. It is aware of the spines of the sea urchin. It also has often tasted the soft underbody of the sea urchin. The puffer swims near, but stays clear of the sea urchin's spines. It begins to blow jets of water around the sea urchin. The urchin tries to hold on to the bottom with its tube feet. But the puffer fish huffs and puffs. It blows the sea urchin loose from the sand. It blows the urchin upside down. Now the sea urchin's soft underside is exposed. The urchin can't use its spines. The puffer moves quickly. It bites into the urchin. The sea urchin loses the battle with death.

☠ In this case, neither animal used its poison.

TOADS

Toads, like other amphibians, can be poisonous. Their poison is usually in glands under the skin. When they're touched or squeezed, the poison comes to the surface. It sometimes has an unpleasant smell and is often bitter-tasting. Dogs and cats that catch toads soon discover that carrying a toad in the mouth is no fun. But the dog that picks up a Marine toad is in big trouble. This huge creature, up to 10 inches long (25.4 centimeters), is highly poisonous and capable of killing the dog that swallows it.

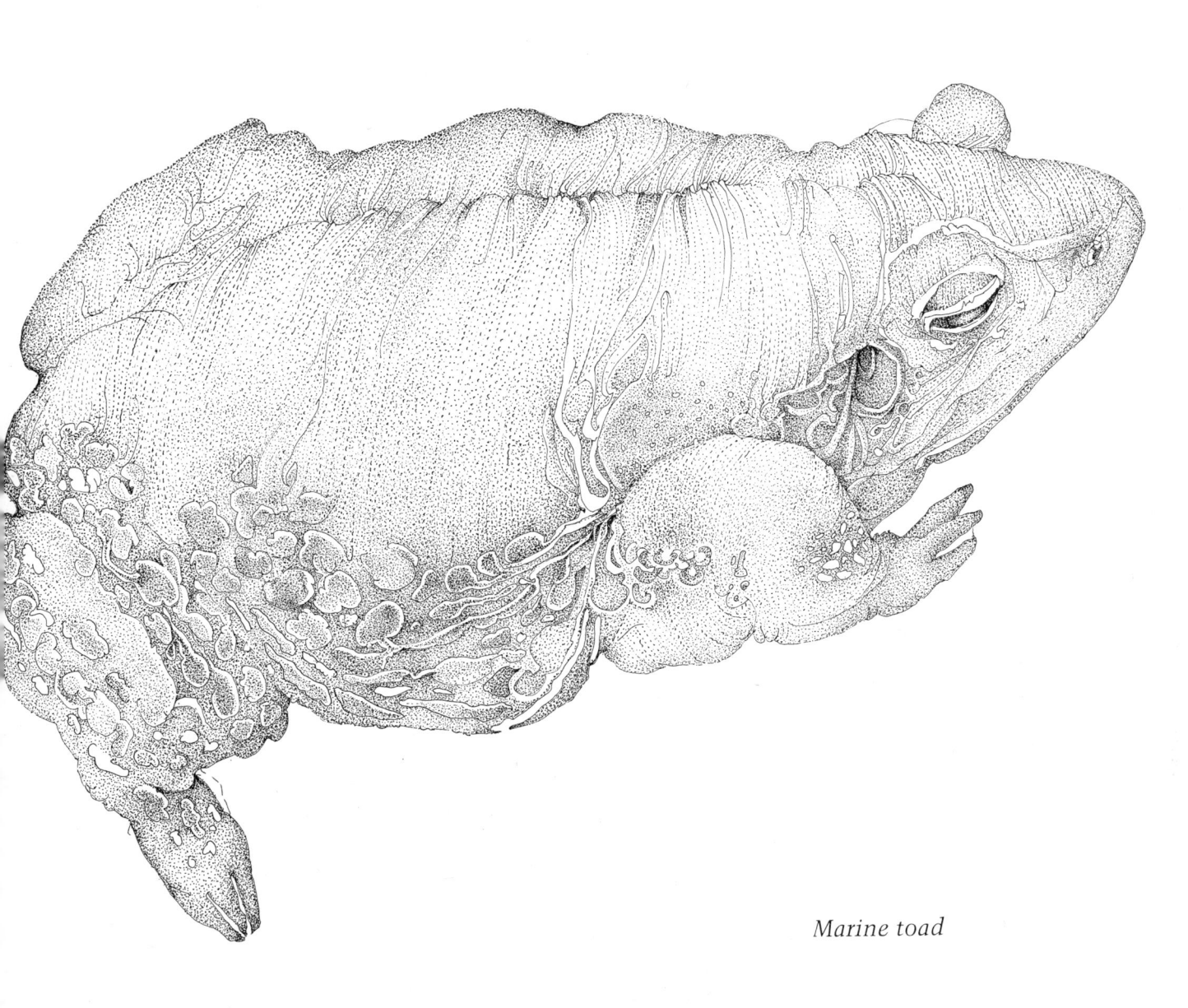

Marine toad

FROGS

Some frogs secrete venom strong enough to kill a person. Many of these frogs belong to a group called poison arrow or poison dart frogs, which are found only in Central and South America. There the Indians collect the poison by spearing the frogs and holding them over a fire. The heat of the fire forces the poison out of the frog's skin. It collects in drops and is scraped off into a pot. The Indians use it to make poison arrows for hunting.

The Kokoi frog is only 1 inch long (2.5 centimeters) and very brightly colored. Its colors may be a warning: "Remember how I look! Stay away! If you eat me, you'll be sorry!" The venom in this frog's skin is ten times more powerful than that of the puffer fish.

☠ One Kokoi frog can produce enough poison for fifty arrowheads.

NEWTS

Newts are aquatic salamanders. They are related to frogs and toads. Not all of them are poisonous, but the crested

Crested newt

newt is one of the poisonous species. The poison is in the back and tail. When something (or someone) squeezes a crested newt, the poison comes out through the skin. It is so bitter that some snakes which eat other newts avoid the crested newt.

The California newt is one small bundle of poison. It has poison in its skin, blood, muscles—even in the eggs it lays. The poison is so strong that an amount so small it can't be seen by the human eye can kill 7,000 mice.

☠   Here's a mystery: Why doesn't the newt poison itself?

MILLIPEDES

These are the animals which are sometimes called thousand-leggers. They are generally small creatures, although a few specimens have been reported of up to 12 inches long (30.5 centimeters). Many species produce poison in glands on the sides of their bodies. The poison is released as liquid or gas and it has a terrible smell. Some millipedes have been known to blind a chicken. The Sequoia millipede, which glows in the dark, can kill a bee with its poison.

Every so often there is a plague of millipedes. Once 65 million of them suddenly appeared on a farm in West Virginia. Their odor was so strong that people got dizzy from it. Cows, horses, and people had to be moved until the millipedes went away.

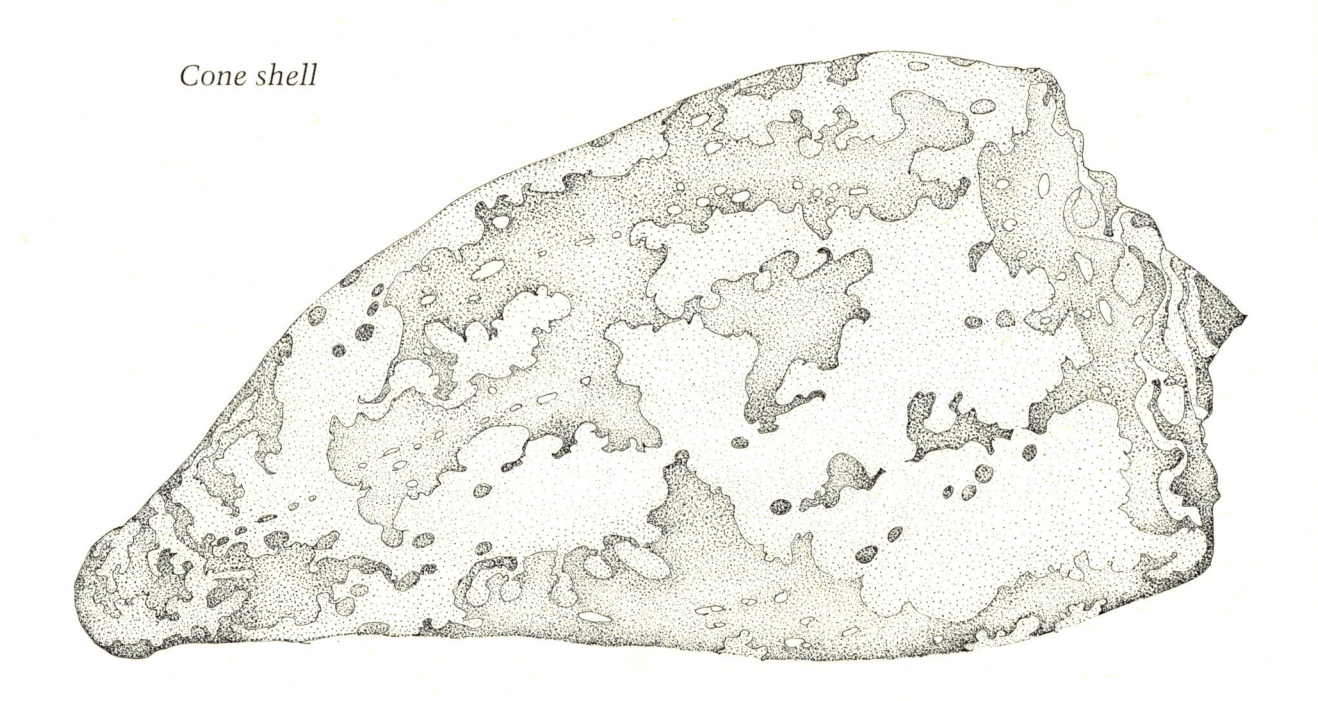

Cone shell

7 · Tooth and Claw

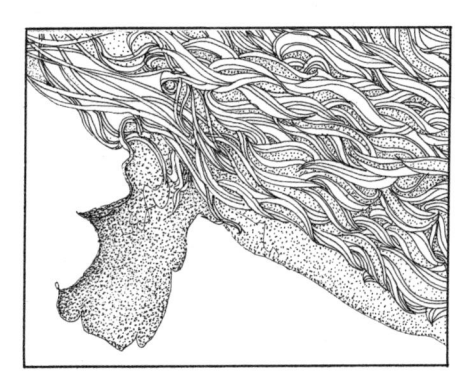

CONE SHELL

Most cone shells are small enough to hold in the palm of your hand. But don't do it! All of them are venomous and some are more deadly than a rattlesnake. Tucked inside every cone shell are poisonous harpoons. The tip of each harpoon is a tooth. Venom is pumped up through the shaft of the harpoon to the tooth.

The cone shell hunts at night. While hunting, its body

is outside the shell. It looks like a small blob of flesh moving along the bottom of the sea. As soon as it makes contact with a sea worm, small fish, or other prey, the cone sends a harpoon out of the open end of its shell. The poison has an instant heart-stopping effect.

☠ Nature's puzzle: Why does the cone shell need such strong poison for such small prey?

SHORT-TAILED SHREW

Shrews are small, furry animals. They live in many parts of the world and are very tough for their small size. The short-tailed shrew is one of a group of shrews that have a venomous bite. The animal's teeth make holes in its victim. Then venom flows into the holes from glands in the shrew's jaws. The prey quickly becomes numb. This powerful weapon makes it possible for the short-tailed shrew to catch and eat prey many times its size. And a good thing, too, for the shrew's survival. Because these little animals have to eat their own weight every twenty-four hours.

☠ If a shrew is without food for as short a time as three hours, it will die.

Duck-billed platypus

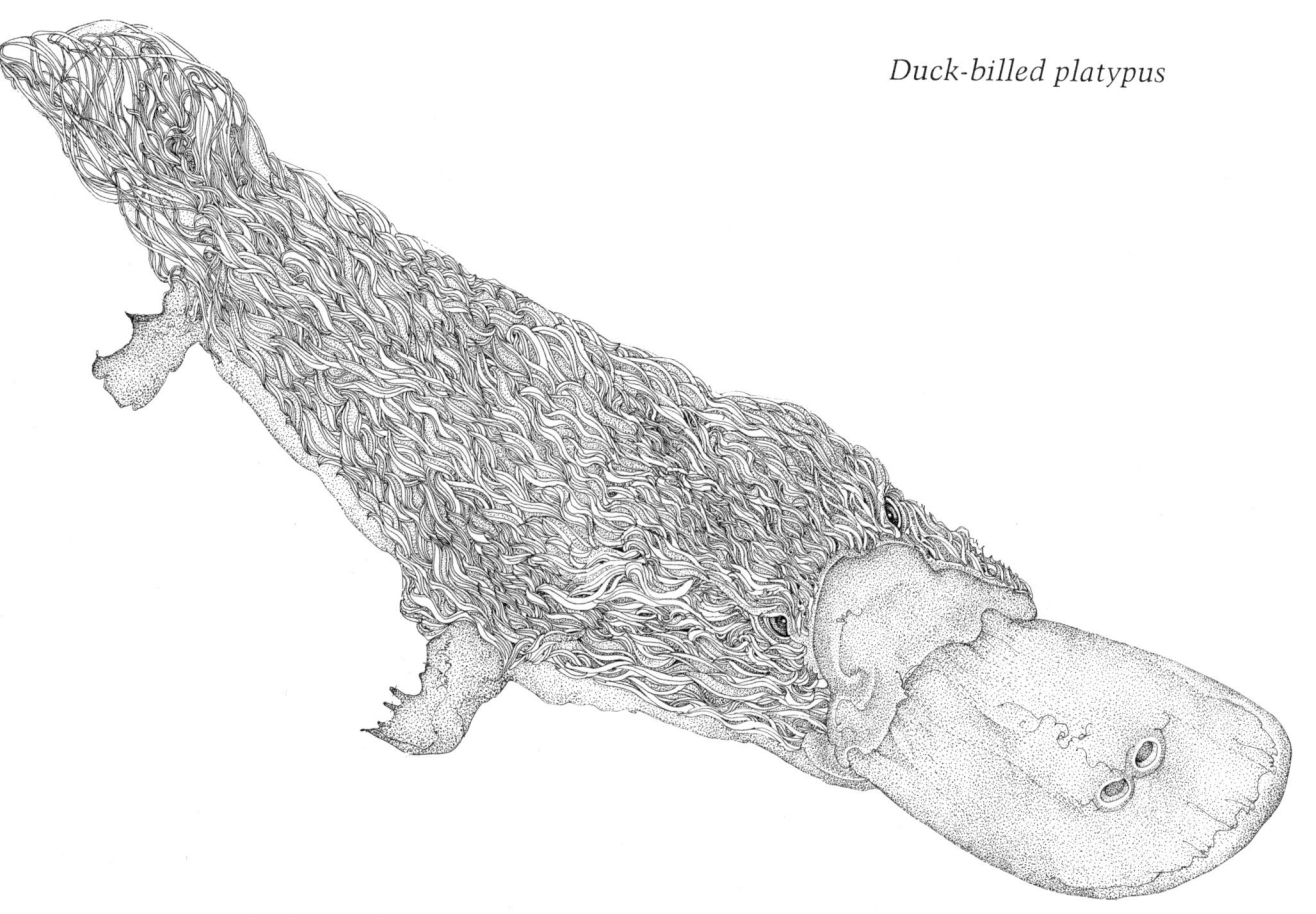

DUCK-BILLED PLATYPUS

The platypus is one of the strangest animals in the world. It lays eggs, like a bird. It has fur, like a seal. It nurses its young, like other mammals. And it has a bill, like a duck.

It also has a poisonous claw on each back foot. What does the platypus use it for? So far, that is still a mystery.

The platypus is only one of the poisonous animals which no longer seem to have any need for their poison. One day scientists may know more about these strange creatures who have weapons they never use.

8 · The Smallest and the Deadliest

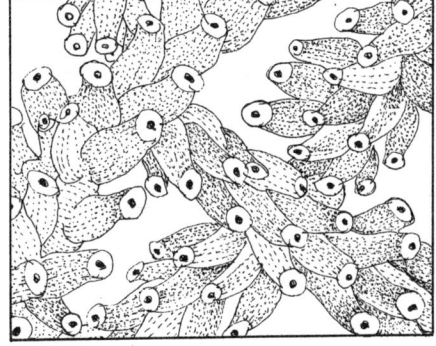

ZOANTHIDS

There is a legend on the island of Hawaii. It says that a witch doctor there put a curse on a fisherman. The man was killed and his body was burned. The ashes were thrown into the sea. From that time on, the story goes, moss grew in the water at that spot. The moss was so poisonous that it could be used to make arrowheads fatal. The location of the place where it grew was kept secret. It was made taboo. Forbidden.

Zoanthids (cluster)

    Recently a group of scientists found the deadly "moss." They discovered that it was actually a colony of tiny sea animals called zoanthids. They are related to corals, and this Hawaiian species contains poison which may be more powerful than that of any other animal shown in this book.

9 · A Final Word

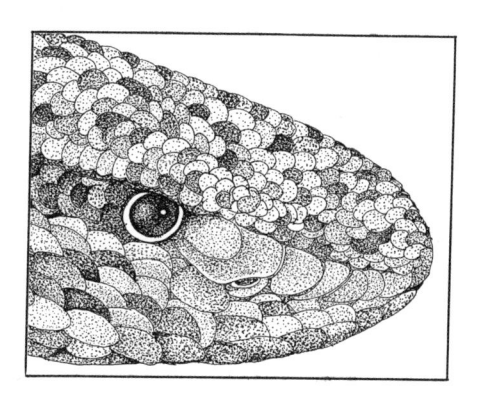

alien ā'lē-ən) 1. A member of another family or people. 2. An outsider.

These are a small number of the poison animals in the world. You will meet most of them only in a zoo or aquarium. But a few of them may live near you. If they do, learn to recognize them. Don't be afraid, but have respect for poisonous creatures. Avoid the places where they are likely to be. Don't hunt them or seek them out. Remember—no poison animal kills for thrills. But it will respond to a disturbing or alien presence. *Don't be one.*

Where to Look Out for Them

ARROW POISON FROGS Central and South America
Family Dendrobatidae

BLACK WIDOW SPIDER Warm parts of the world
Lactrodectus mactans

BOOMSLANG Africa
Dispholidus typus

BROWN RECLUSE SPIDER South Central states of U.S.
Loxosceles reclusus

BUSHMASTER *Lachesis muta*	Northern S. America, Panama, parts of Costa Rica
COBRA, INDIAN *Naja naja*	Africa, tropical Asia
COBRA, SPITTING *Naja nigricollis*	Africa
CONE SHELL *Conus geographus* and others	Western Atlantic and other tropical waters
COPPERHEAD *Ancistrodon contortrix*	Eastern half of U.S. to Gulf states
COTTONMOUTH WATER MOCCASIN *Ancistrodon piscivorus*	Southern Atlantic and Gulf states of U.S.
DUCK-BILLED PLATYPUS *Ornithorhynchus anatinus*	Australia
GIANT WATER BUG *Family Belostomidae*	North America
HONEYBEES, BUMBLEBEES AND OTHERS *Order Hymenoptera*	All over the world

MARINE TOAD *Bufo marinus*	Texas, Florida, Central America, Caribbean, Hawaii
MILLIPEDES *Class Diplopoda*	Widely distributed
NEWT, CALIFORNIA *Taricha torosa*	California
NEWT, CRESTED *Triturus cristatus*	Europe, Asia, N. Africa, N. America
PORTUGUESE MAN-OF-WAR *Family Physalidae*	Warm waters worldwide
PUFFER FISH *Tetraodon hispidus* and others	Tropical seas and large rivers like the Congo and Nile
SCORPION *Centruroides sculpturatus* *C. getschil* and others	Warm places throughout the world
SCORPIONFISH *Family Scorpaenidae*	Warm tropical waters, especially Japan, the Philippines and northern Australian coasts

SEA ANEMONE *Order Actinaria*	World-wide seas
SEA URCHIN *Toxopneustes pileolus* and others	World-wide seas
SEA WASPS *Cubomedusae of Class Scyphoza*	Temperate seas around U.S and elsewhere
SHORT-TAILED SHREW *Blarina brevicana*	North America
SIDEWINDER *Crotalus cerastes*	Parts of southwest U.S. and Mexico
STINGRAY *Dasyatis americana* and others	Shallow temperate and tropical seas. Some rivers.
STONEFISH *Scorpaeonopsis gibbosa*	Red Sea, Indian Ocean, North coast of Western Australia
TARANTULA *Theraphosidae family*	North America
TARANTULA HAWK and Other Wasps *Order Hymenoptera*	Widely distributed

TIMBER RATTLESNAKE   Eastern U.S.
Crotalus horridus

ZEBRA FISH   Widely distributed throughout
Pterois volitans   Indo-Pacific

J

A23678 ✓

591.6   Brenner, Barbara
B   Beware! these animals are poison. Illus.
        by Jim Spanfeller. New York, Coward, McCann
        & Geoghegan [c1979]
              63 p.

            1. Poisonous Animals. I. Spanfeller,
        James J., Illus. II. Title.